DUCKS AND GEESE

A Guide to Management

DUCKS
AND GEESE

A Guide to Management

TOM BARTLETT

THE CROWOOD PRESS

First published in 1986 by
The Crowood Press Ltd
Ramsbury, Marlborough
Wiltshire SN8 2HR

www.crowood.com

This impression 2004

British Library Cataloguing-in-Publication Data

Bartlett, Tom
  Ducks and geese: a guide to management,
  1. Ducks   2. Geese
  1. Title
  636.5´97   SF505

ISBN 1 85223 650 7

Acknowledgements

Photographs by Tom Bartlett

Line illustrations by Vanetta Joffe

Typeset by Alacrity Phototypesetters, Weston-super-Mare

Printed and bound in Great Britain by CPI Bath

Contents

Foreword

Interest in keeping domestic waterfowl has never been greater than it is today. Following two decades of decline in the fifties and sixties, the early seventies saw a great revival of enthusiasm for the variety of domestic breeds, encouraged by imported strains from top American breeders.

At about the same time, the British Waterfowl Association started to promote domestic waterfowl, reaching out to an ever widening audience. The biggest promotion undertaken was the Village Pond Campaign. This was an unqualified success which involved people from many walks of life. It resulted in ponds being cleaned up nationwide, and these became homes for ducks and geese. In addition, by sponsoring waterfowl classes at local and national shows, the BWA has brought home the value of conserving rare breeds. This has been enhanced by their presence at game and county fairs, and smaller local events.

Tom Bartlett at Folly Farm has developed a fine collection of waterfowl. By opening the Poultry to the public, he has done a great deal to promote awareness of the beauty of our widely varying domestic breeds. This is the only collection open to the public which concentrates on these domestic forms, and Tom is to be congratulated on his enterprise. His enthusiasm for his subject has brought him great success at the top waterfowl shows; and now his knowledge and experience of livestock, which come from a lifetime's association with domestic animals, are written down for us all to enjoy. Although this book is intended for those just starting out, it will, I know, appeal to everyone interested in keeping domestic waterfowl.

Christopher Marler
British Waterfowl Association

Introduction

Folly Farm Poultry, near Bourton-on-the-Water, began as a private collection of rare breeds of domestic fowl and over the years has expanded, becoming one of the largest and most interesting collections in the country. During this time it has helped to perpetuate some of the breeds which had almost disappeared, especially in the field of domestic waterfowl.

In the drought of 1976 three lakes were constructed on part of the farm for the purpose of water conservation. These also provided a suitable area and, therefore, an opportunity for expansion in the breeding and rearing of domestic waterfowl. Recently this collection has been opened to the public in a hope that, having seen the wide variety of waterfowl displayed in their natural environment, people will realise how easy it is to keep them and maybe understand how they too can participate in the preservation of species which might otherwise become extinct.

The success of this venture prompted me to write this book, illustrated with photographs of the stock at Folly Farm, and hopefully it will answer some of the many questions put to me every day, by the constant stream of visitors.

Based on our experiences at Folly Farm, the methods described here are meant as a practical introduction, to be used mainly as a guide for the beginner wishing to keep waterfowl. There may well be other enterprises which are equally successful but have achieved this with different facilities.

My thanks must go to my wife, Diana, because without her continued support and added knowledge of all aspects of keeping waterfowl, I would have found it impossible to establish Folly Farm as the successful enterprise that it has become. Also, thanks to Peter, my son, who has the understanding to manage and care for the stock in our absence.

Introduction

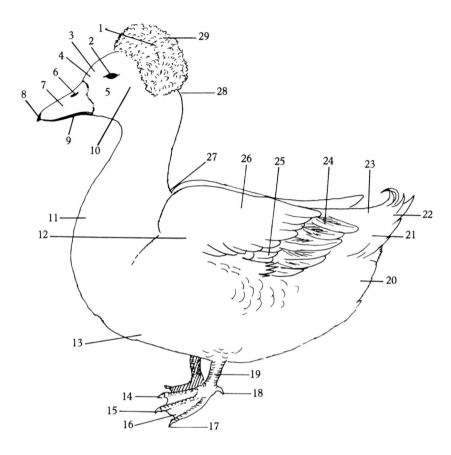

KEY

1	Crown	11	Breast	21	Upper tail coverts
2	Iris	12	Wing coverts	22	Tail feathers
3	Forehead	13	Belly	23	Rump
4	Face or lores	14	Inner toe	24	Primary feathers
5	Malar region	15	Middle toe	25	Speculum
6	Nostril	16	Webbing	26	Secondary feathers
7	Upper mandible	17	Outer toe	27	Mantle
8	Nail	18	Hind toe	28	Nape
9	Lower mandible	19	Shank or tarsus	29	Crest
10	Ear coverts	20	Under tail coverts		

Points of a duck.

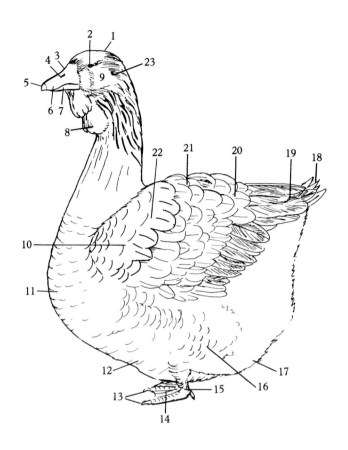

KEY

1	Crown	9	Face or lores	17	Paunch
2	Eye	10	Wing bow	18	Tail feathers
3	Culmen	11	Breast	19	Primary feathers
4	Nostril	12	Keel	20	Secondary feathers
5	Nail	13	Toes	21	Scapula
6	Upper mandible	14	Webbing	22	Shoulder
7	Lower mandible	15	Shank or tarsus	23	Ear
8	Dewlap	16	Thigh coverts		

Points of a goose.

1 Keeping Ducks and Geese for Pleasure

When I was a young man, the object of keeping waterfowl was, of course, food production. If your duck did not lay enough eggs to earn its keep, it was soon for the pot! If a bird meant for Sunday lunch did not put on sufficient weight over a period of time, you would have to look for a different strain. In those pre-war days every farm kept its ducks, chickens, turkeys and geese. Some were for consumption on the farm and the rest were taken to the local butcher where the farmer hoped to receive a fair return for his efforts.

Of course, nothing is ever static for long. The forties brought war, so we had to choose the kind of bird which would make best use of the food available. Our ducks, which were the laying variety, were fed on boiled fish offal from the local fishmonger and barley meal. This gave a rather unpleasant fishy flavour to the eggs, but nobody bothered in those days – a 'fishy' egg was far better than no egg at all! Ducks thrived on this diet and laid well.

Following the War, even greater changes took place. The small flocks of free-range chickens, ducks and geese that used to roam the farmyard practically disappeared. They were considered inefficient and the word 'agri-business' crept into the farmer's vocabulary. There was a race by the large companies to produce chicks by the thousand to supply the large battery units which were springing up all over the country. I consider it fortunate that ducks and geese were not ideal for these factory-like conditions. In time the small flocks of duck, chicken and geese almost disappeared from the farmyard scene and, if it had not been for some of the smaller farms in more remote parts of the country (such as Cornwall and Wales) who took pride in our pure breeds and kept them on, these breeds may well have become extinct.

Keeping Ducks and Geese for Pleasure

It was not until the late sixties and early seventies that interest in our pure breeds began to grow once again. To begin with perhaps for some it was a nostalgic desire to recreate that happy farmyard scene from their youth, but when waterfowl really began to exert their appeal again in the early seventies, the attraction was quite different. This time it was for pleasure, and the eggs they laid were a welcome bonus.

Keeping ducks for pleasure does not only exist in the farming community but with anyone who has a garden large enough to support a pair or trio. When people see a beautiful bed of flowers they are tempted to stop and admire them, but if there is a pool with a few ducks they will stay for much longer, fascinated by the antics, as the duck asserts its appeal. This sight tends to make people slow down, such an essential part of life in these fast-moving times.

Scarcity will always inject interest, as well as value. Look how the Rare Breeds Survival Trust has grown from strength to strength over the last ten years. The British Waterfowl Association has never been so popular as it is today. In our lifetime domestic ducks and geese, and their keepers, have had to continually adjust to change, but their present role now seems well established. No longer do the economics of food production apply: their future existence is now as a hobby, and one which is not only popular in this country but is being established worldwide.

Ducks and geese have characteristics which seem to appeal to most people. They come in all shapes, sizes and colours and all have one thing in common – the ability to adapt easily to human company. With their unique, captivating appeal, I firmly believe that interest in ducks and geese will continue for many years to come.

2 Keeping Ducks

First of all, you must consider what conditions you have for keeping ducks. It is a great mistake to rush into buying the birds and only think of their requirements afterwards. Essential to successful keeping is that your ducks become tame, happy and contented, so never try to keep too many in a confined area.

One of the big advantages of keeping ducks is that they do not require elaborate housing with perches, nesting boxes and special ventilation. Another is that, with the exception of the small ducks, most of the domestic breeds do not fly, so 2in (5cm) mesh wire netting of about 3ft (1m) high will confine the birds and keep them where you wish. Providing you don't have too many, you can let them run in the garden – ducks do not scratch and will do very little damage, and they will very quickly rid the garden of slugs and snails. If you are attracted to the small ducks which do fly, by clipping a few of the pinion feathers of one wing, you can keep them grounded until they grow their new feathers. By that time, if their environment is right most of the small ducks will be happy to stay there. Ducks enjoy swimming in water but it is not absolutely essential. Providing there is enough water for them to immerse their heads and dabble, they will survive and perform quite satisfactorily.

All domestic ducks will, of course, interbreed, so if you are planning on keeping more than one strain, it is important to have a separate pen and house for each individual variety. You will need to consider the minimum area required for the small or large breeds. I would not recommend less than 6yd (6m) square for the small and at least double this area for the large varieties. If the land is well drained and dry, you may carry up to four or five birds in this area, but I stress that this is the minimum required. If you are fortunate enough to have additional land on which they can range then, after you have bred sufficient birds for your needs, you can let them run as a flock for the rest of the year.

THE DUCK HOUSE

The kind of house required for duck is very simple. No perches or nest boxes are needed and there is no point in having windows for light, as ducks do not use the house in daytime. They do, however, need adequate ventilation. Most important is to ensure that your duck house has a dry floor.

Flooring

You may well have a disused building which would be quite simple to convert, but if you are making the house, at Folly Farm we have found that a simple apex construction is perfectly acceptable. About 3ft (1m) square at the base is sufficient for a trio of small duck, with about a third as much again for the larger breeds. We fix ½in (1.25cm) netting or weldmesh to the bottom of the structure, protruding about a foot outside the perimeter of the house to stop predators, such as rats or foxes, scratching their way into the house from underneath. This is

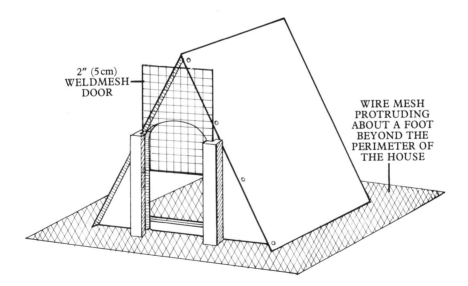

2″ (5 cm)
WELDMESH
DOOR

WIRE MESH
PROTRUDING
ABOUT A FOOT
BEYOND THE
PERIMETER OF
THE HOUSE

Fig 1 A simple but adequate duck house.

13

The duck house need only be very simple but must have an adequate opening to provide ventilation.

covered with a thick bed of clean, dry straw. This bedding will need twice weekly attention in wet weather, which entails peeling off the 'patted down' top layer, usually about two inches, and replacing it with another armful of clean, dry straw. We have found this method very effective at Folly Farm which has easy draining soil.

If the land on which you keep your birds is wet or has poor drainage, you will need a floor in the house. As ducks do not take kindly to wire or slatted floors, a wooden one would be the most satisfactory for the average small house. Make sure, however, that the wooden floor is a good 6-8in (15-20cm) from the ground: if it is too close, it becomes an ideal harbouring place for rats. A raised house is more difficult for a duck to enter and can cause panic, so a running board is required. Make this the same width as the entrance and well sparred because a board made slippery in wet or icy weather can cause chaos and put the ducks off lay for days.

Ventilation

Ventilation is very important as ducks tend to be claustrophobic. We have found that making the door of weldmesh (about 1½in (4cm) mesh, measuring 2ft × 2½ft (0.6m × 0.75m)) is an excellent way of overcoming what might otherwise be a problem. Unlike chicken, ducks do not like 'pop holes', they like to go in and out of their house in a rush.

If the house is agreeable to them, it is a much easier task to shut them up at night. An unsuitable house will lead to panic and your duck will never perform satisfactorily if there is any form of stress.

WATER

Swimming water is not absolutely essential for waterfowl, yet it pays to provide it wherever possible. However, as long as there is a receptacle containing enough water for the birds to immerse their heads in, a satisfactory performance can be achieved. In 1976 and 1984, years of extreme drought for our climate, the springs that supply the pools and lakes at Folly Farm dried up. We had to rely solely on the mains water. Five gallon plastic drums were split in two to make water butts which were filled with fresh water every day. The birds did not seem to miss their swimming water and on both occasions came through in excellent health with, as far as we could tell, no detrimental effects in any way. However, as a general rule the more water they have, the happier they seem to be.

Plenty of swimming water is a great help to feathering, so if you wish to show your birds abundant clean water is advantageous. One of the greatest assets in the preparation for showing is a completely clean pen - every effort must be made to avoid mud. Naturally, this difficulty will be greatest around the edge of the pool. If you are fortunate enough to have natural swimming water, either a stream or a pool, make sure that the bank is not too steep. If the bird has to flap its wings to help itself up the bank, especially if the soil is clay or sticky mud, it will discolour the pinion feathers and ruin it for showing. A gentle slope giving the duck ease of access to the water, and out of it, is very important.

If you wish to make an artificial pool, it is not difficult. I much prefer

Ducks enjoy swimming in water although it is not essential for them.

A halved five gallon drum is a satisfactory alternative when water supplies are limited.

16

*All ducks benefit greatly from having a shaded spot
to rest in during hot weather.*

to purchase a fibreglass mould, rather than using plastic sheeting, which you can obtain from any garden centre. They are made in many shapes, varieties and depths and are easy to install. It is the edge of the pool that can become muddy and needs the most attention. You can surround the pool with gravel or pebbles but make sure the material you use is rounded and not sharp as this could damage the bird's feet. The gravel should extend about two feet from the pool edge and be fairly deep so that the rain will continually wash it off and keep it clean. If you do decide to have a fibreglass pool, remember that the sides are very slippery. As long as the pool is kept full of water the duck can get out easily, but if the level is allowed to drop the duck will experience great difficulty. Putting some bricks at one end is an easy solution to the duck's problem.

WEATHER CONDITIONS

In the summer months static water can turn green very quickly. A plug at the bottom of the pool with a soak-away under it are strongly recommended, so that the pool may be drained easily. In hot weather

Fast-growing trees may be planted to provide shelter from cold winds.

the water will need changing at least once a week. During the winter months the pool may freeze over from time to time but your ducks still need to drink and immerse their heads. Supply them with a small separate receptacle for this purpose and check that it is full at least twice a day.

Ducks don't mind the cold providing they have plenty of food and water: this should be provided twice a day. As ducks lose most of their body heat through their feet and legs, litter a little straw on the ground when the weather is at its coldest and the birds will be quite happy. At this point, I must give a warning to those with larger lakes and pools. Realising that the ducks need water, you may be tempted to go out on the lake and punch a hole in the ice for the ducks to drink. They will soon find the water and slip in for a swim, but as they are surrounded by an edge of ice it will be difficult, if not impossible, for them to scramble out, and having become exhausted by their efforts they will soon succumb to the cold water. Many ducks are lost in this way every year.

Another weather condition that ducks dislike is high wind, so they need a sheet of tin or a bale of straw behind which they can shelter. Ducks also benefit greatly, in very hot weather, from the provision of plenty of shade.

3 Breeds of Duck

SMALL BREEDS

If you have limited space, you may wish to select one of the three small breeds. These are popular because they are miniature which, for some reason, has a particular fascination of its own. Miniature horses, Shetland ponies, pigmy goats and bantam fowl are well loved around the world, and often breeders of waterfowl come under the spell of breeding miniature ducks. First, you should consult the *British Waterfowl Standards* to establish firmly in your mind what are the Standard characteristics to be achieved.

Call Duck

During and after the War, when the breeding and rearing of British waterfowl was at its lowest ebb and our best breeding strains had dwindled to a pitiful number, Christopher Marler, one of our leading breeders, realised the importance of bringing new blood into the country and imported numerous domestic breeds from America, including the white Call duck. This outstandingly attractive bird must have drawn more people to join the ranks of waterfowl breeders than any other single species. There is a wide choice of colours from which to breed, the white, at the moment, being the most popular. This bird should be pure white, free from any tinge of yellow, with a bright orange-yellow bill and dark eyes. There are also the brown, pied, blue/fawn and silver Call ducks, all of which have their exact markings described in the *British Waterfowl Standards* book.

Whichever colour you choose to breed, they all have the one outstanding feature and that is the shape of the head. This should be rounded, with a short, broad bill: a long narrow bill looks very unattractive in any Call. The body must be small and compact and, as the Standard dictates, the drake should weigh 20-24oz (570-680g) and

Popular white Call duck, improved with American blood.

Fawn Call duck of the Old English type – note the longer bill and more racy appearance.

The silver Call duck (right).

the duck 16-20oz (450-570g). When selecting breeding stock, people are often tempted to choose the smaller birds, but the Standard weight, if well proportioned, will certainly be a better breeder.

This bird was originally bred by British breeders because of its loud call and was used as a decoy for 'calling in' wild varieties for shooting. Nowadays, however, this call can be a problem and if you are considering keeping this breed, make sure that your close neighbours don't mind the sound that these little ducks make. I must say that I find the chatter of the little females rather appealing but, as it can be fairly constant in spring and autumn, some people may become irritated by it.

There is an added attraction to keeping the Call duck in that they are quite capable of hatching their own young and rear them with admirable skill and devotion, but one of their most endearing characteristics is their friendliness which can make them good companions. Some people will keep them as pets, but they also provide a challenge to the more serious breeder as new colours and even crested varieties are introduced and need perfecting. I feel absolutely sure that these charming little birds have a wonderful future.

To be successful at keeping and breeding Call duck, you will need an area of at least 6yd (6m) square, for they must have plenty of exercise; also, given the chance, they are excellent foragers. Your first task is to find a pair or trio of the best quality provided by a reputable breeder. Remember that top quality birds cost no more to feed than a poor representative of the breed. Make sure that one wing is clipped or that they have been previously pinioned. A house about 3ft (1m) square is ample for a pair or trio.

Black East Indian

This duck is strikingly beautiful with its lustrous green sheen, especially strong on the head, neck and back, and its slate black bill. Its nature is slightly more nervous than the Call duck but they are very hardy and

The Black East Indian with its lustrous sheen makes an attractive contrast to the other small varieties.

excellent foragers, benefiting greatly from being allowed to range free. There is, however, a problem in that after a few years of breeding white feathers start to appear. The older the birds are, the whiter they will become. There is no harm in breeding from these females as their offspring should be true to colour, but some strains turn white quicker than others, so select those that keep their colour longest for the breeding pen.

Miniature Silver Appleyard

This small duck was bred originally by one of the finest waterfowl breeders of our time, Reginald Appleyard. It is really a small version of the large Silver Appleyard, but in recent years some have strayed from the original Standard and many have shown characteristics of the Call duck. This bird should be an exact replica, in all but size, of its larger

Miniature Silver Appleyards – the only miniature breed with a larger counterpart.

counterpart, with a small, neat, slender head and medium length bill. It is the only miniature of a larger breed in the world of waterfowl. The better the standardised version of the breed there is on display at the shows throughout the country, the more accurate will be the visual impression of the true Miniature Silver Appleyard carried in the minds of breeders everywhere. For the more serious breeder, this bird certainly still gives plenty of opportunity for him to prove his expertise.

I find these birds the best layers of the small breeds. They are good sitters and make very caring mothers. They have a very quiet nature – nothing seems to worry them – and they adapt easily to new conditions. This is arguably the most suitable duck for those with limited space.

LIGHT BREEDS

Indian Runners

The Indian Runner ducks were first introduced to Britain from Malay by a sea captain who brought them into the Border country and distributed them amongst his farmer friends. They are an extremely active breed and, if you are able to give the Runner free range, it is an ideal bird for that purpose, wandering far and wide, picking up most of its own living for certain seasons of the year in the form of natural foods. In my view, Runners are the most graceful and elegant of all ducks and with their fine, upright carriage, never fail to capture the interest of an observer. A dozen or so, running in a long line on their way home in the evening is an unforgettable sight.

These ducks should be very slim with an upright stance. They will generally stand at an angle of about 80° to the ground, but when startled will be almost perpendicular. The total length of the drake should be 28-32in (71-81cm) and the duck 24-28in (61-71cm). They have a very long neck but this should not be more than one third of the total length of the bird. The body should be slim, elongated and well rounded, somewhat similar to the shape of a hock bottle – with legs! The wings should be well tucked in, with the tips of the long flight feathers just 'kissing' at the back. When startled and standing upright, the tail should be straight in line with the back. The head should be lean and streamlined.

Runner ducks need less water than most varieties: quite satisfactory

*A group of white Indian Runners, showing the ideal
shape and stance for the breed.*

The trout-coloured Indian Runner.

25

Breeds of Duck

The fawn and white Runner.

results can be achieved providing there is a tub in which they can immerse their heads and dabble. Where swimming water is available they do not spend much time on it but, of course, swimming water does help feathering. They are first class layers, and it was the Runner duck's egg laying capabilities that brought it fame. Soon their distribution was countrywide. Mrs Campbell of Uley in Gloucestershire used the wild Mallard, Rouens and the fawn and white Runner to produce the Khaki Campbell, and there can be no doubt that the excellent egg laying abilities of her Khaki Campbell were inherited from the fawn and white Runner.

There are many colours of Runners, and a mixture of all these in one flock form a very attractive sight. It must be remembered, however, that all these colours will interbreed, so it is vital to separate them at breeding time. A great deal of time, expense and expertise has been spent to produce these colours and markings; it is a shame that every year so many of these superb species are lost in cross-breeding.

A typical exhibition type of Khaki Campbell.

This breed has a more nervous disposition than the heavier types of duck so there is a greater need to provide the correct environment and to handle them carefully, especially if you wish them to present themselves well at a show. This way, the Indian Runner *can* become very quiet. In the breeding pen, five ducks can be run with one drake quite success-fully for up to five seasons, but it is worthwhile changing the drake every two years.

Khaki Campbell

The Khaki Campbell is probably best known for its prolific egg laying ability. The pearly white eggs are a feature of the breed and should never be green. When Mrs Campbell created this breed she wished it to become the top commercial egg producer rather than an exhibition breed. She obviously felt that breeders for exhibition would give too much emphasis to the size and colour of the bird and may lose sight of the commercial purpose for which she originally intended them. However, in the exhibition pen today you will see beautifully coloured, well-proportioned Campbells. Maybe they have lost some of their egg laying prowess in the process of breeding for show, but nowadays the perpetuation of these breeds depends far more on their beauty than their effectiveness as egg producers. In any case, a pen of five duck will still produce more eggs than most householders require.

White Campbell

The White Campbell is a white sport from the Khaki. If this bird is of genuine, full-blooded Campbell stock, it is excellent, but many of this breed are of doubtful origin and there can only be a few breeders left who could genuinely claim to have pure, White Campbell stock.

There is also the Dark Campbell which was created by Mr H. R. S. Humphries of Devon to make sex-linkage in ducks possible.

*The White Campbell used to be popular but is
not often seen today.*

Crested Duck

The Crested duck may be bred in many colours but improving the crest commands more attention. If you have a well-made bird of medium length and width, weighing 6-7lb (2.7-3kg) then you can concentrate your efforts on the crest. The parent stock need not carry large crests to still produce some fine exhibition specimens. The crest should be round on both sides and on top. It should be large, well-balanced and firmly anchored. Avoid those birds in which the crest appears off-centre or lopped to one side. Also to be avoided are split crests and those set too far back on the head or even the neck. These are faults that are very difficult to breed out.

The big advantage in breeding Crested ducks is that you can see how good the crest will be immediately they have hatched. As there are going to be a percentage that will hatch with no crest at all, you can assess the quality of your ducklings at a day old. The breeding pen may well consist of five or six ducks to one drake.

White and coloured Crested ducks.

Breeds of Duck

White Crèsted duck showing well-balanced crests.

The Welsh Harlequin.

Welsh Harlequin

This is a utility duck which can be very attractive in appearance and is an excellent layer, but for the show pen I have found it a rather disappointing breed because of the wide variation in colour and type. Nevertheless, because of the interest that surrounds this breed, I am sure that breeders will achieve a greater uniformity and come closer to the written Standard than we have seen in the show pen in recent years.

HEAVY BREEDS

Aylesbury

Many people have heard of the Aylesbury duck but few know what a magnificent bird it really is. In recent years, I believe it is fair to say that any white duck has been called an Aylesbury whether or not it is the true breed, either in type or characteristics. Many are simply mongrels with little or no Aylesbury blood.

When originally developed the Aylesbury duck was considered by breeders and poulterers to be one of the best breeds for the table. It matured quickly, gaining considerable size and weight at an early age, and the flavour and quality of its flesh were unsurpassed. Vast numbers were bred and raised in the Vale of Aylesbury, from where the breed originated and took its name. These birds supplied the nearby London market and, because of this tremendous turnover of stock, they became very inbred. Breeders then began to cross them with the Pekin and produced a bird with much hybrid vigour which led to the popularity of the Pekin × Aylesbury cross, but to the loss of much of the pure Aylesbury strain.

What a majestic bird this is, with its long, deep body, a straight keel from breast to stern that is practically parallel to the ground. The plumage should be pure white – any blemishes disclose an impurity in the blood. Except for the greater size and the curled tail feathers of the drake, the general characteristics are similar in both sexes. The bills should be broad, long and a light pink colour – typical of the true Aylesbury. A yellow bill definitely indicates an impurity in the stock.

Since Aylesburys are bred for meat, they are only moderate layers.

Aylesbury duck showing true Aylesbury type.

They seldom start laying before the middle of February and usually finish in mid-June when they go into a moult. Owing to their great size, fertility can be a problem. It has been found essential to select an active drake, not too heavy and known to have come from fertile stock. He should be mated with two heavier type ducks with good keels. To ensure effective mating, at Folly Farm we have found that it is essential for them to have a pool of water about a foot deep. This encourages the birds to mate and helps the heavier birds to balance whilst doing so.

The breeding of the true Aylesbury has given me great pleasure. It is intelligent, has plenty of character and is a most rewarding bird to show. If it should be a winner of its class it also has a good chance of becoming an overall show champion.

Rouens

This stately bird undoubtedly came from France, and when it was first brought to England it was developed for its table properties and used in cross-breeding table ducks. Today, however, as with most of our breeds, it is bred for its beauty in plumage and markings. It is a huge

The beautiful Rouens with the similar blocky shape of the Aylesbury.

bird with striking colour patterns in both sexes. It is not easily upset and will settle down quickly in the show pen.

The body should be long, deep and broad, and has the square 'block' shape of the Aylesbury, also being of a similar weight. Large adult drakes will sometimes weigh in excess of 10lb (4.5kg). The keel, also like that of the Aylesbury, should be straight along the bottom, parallel to the ground and nearly touching it, but the back of the Rouens should be quite decidedly arched.

In recent years I have noticed that some of these very large ducks lack colour, so study the Standard of perfection carefully and make sure that your stock have correct markings and that their colour is strong. The Rouens will depend on colour for its future popularity.

The breeding pen should be similar to that for the Aylesbury, one drake to two ducks, and to encourage and help them to mate, they need a pool of water. The Rouens do lay quite well even though they are a meat breed. Room to exercise is important and it is a good idea to make them travel for water and food, otherwise they can become lazy and too fat, very often making them less fertile.

Breeds of Duck

Saxony

This breed originated in Germany and has been gaining popularity at the shows during the last few years. It is a dual purpose duck and although it is a solid, meaty bird, you can expect at least 150 eggs per year.

The colour blending is beautiful. The drake, when in full colour, has a pigeon blue head extending as far as the white neck ring on the lower part of the neck. The breast is rusty red with light silver lacing. The tail and wing feathers are an oatmeal colour with blue wing bars. The duck is a lovely shade of buff with an elegant white eye line, while her wing bars and tail are shaded light blue. In both sexes the bill is yellow. On the whole, they breed very true to Standard and a dozen or so of these duck on a pool cannot fail to attract attention. One drake to four ducks seems to be a satisfactory mating.

Cayuga

This breed takes its name from Lake Cayuga in America. They found their way to this country in the year 1871 but early specimens were not very brilliant in colour so British breeders set about improving their colour as well as size and type. They have now established a very beautiful breed, especially the drakes, with their iridescent beetle green feathering. Great care must be taken, however, when choosing birds for the breeding pen, to avoid any purple or brown tinge. Also make sure that there are no birds with white feathers. After the first year, the Cayuga is inclined to show a white feather or two and these will increase with each yearly moult. This, of course, is not a sign of impurity in the breed but some strains will produce these white feathers quicker than others, so it is important to select birds that hold their colour the longest for your breeding stock. I have seen a breeding pen almost white with age, but still producing perfectly coloured progeny.

The Cayuga is a very hardy breed, an excellent forager and a moderately good layer. Do not be surprised, however, if the first eggs that they lay are completely black! They look like black marble and this is a good indication that the strain has plenty of pigment and that the resulting birds will be of a good colour. After the first dozen or so eggs, the colour will become a light grey or white, which is more acceptable if

The pigeon-blue head and rusty red breast of the Saxony drake (in the foreground) shows a good example of this colourful breed.

With its brilliant beetle-green sheen the Cayuga never fails to catch the visitor's eye.

you wish to eat them. A black egg may be a little off-putting even though the content is just the same as one a conventional colour. One drake to five ducks is a good mating.

Blue Swedish

The Blue Swedish duck is gradually gaining popularity. They have always been admired for their striking appearance – a bird with rich, well-laced, clear blue colour is bound to leave a lasting impression. It is not everyone's breed, however, for you must have plenty of room.

The way to select good exhibition stock is to breed plenty and choose the very best for the show pen. To obtain the correct white markings on the chest and wings is a challenge. The Blue Swedish will also produce a certain percentage of black or silver offspring as well as blue. An advantage in rearing the Swedish ducklings is that they grow fast and make excellent table birds, so there is no difficulty in disposing of the stock that do not make the grade. Despite the problems of breeding good exhibition stock, the Blue Swedish will continue to attract breeders who enjoy the ultimate challenge. One drake to four ducks in the breeding pen is recommended.

The markings of the Blue Swedish duck present a challenge to breeders wishing to achieve the perfect Standard.

Silver Appleyard

This breed takes its name from Reginald Appleyard, remembered as one of the most respected judges and breeders of our domestic fowl. It is a breed that was lost and found. For many years at Folly Farm we have kept Silver Appleyard ducks, both the large and miniature varieties. We ran into difficulties when we discovered that the Standard had either been lost or mislaid. I was fortunate enough to locate Reginald Appleyard's daughter, who helped us by producing a painting of a pair of Mr Appleyard's winning birds by the artist Wippel, dated 1947. A comparison with the current birds showed clearly the drift from the original Standard and disclosed the missing characteristics towards which we should have been breeding. From then on we were able and lucky enough to make good and rapid progress towards the true Standard. To quote Mr Appleyard's brochure, written in the 1940s, the object was to make '. . . an effort to breed and make a beautiful breed of duck, with a combination of size, beauty and lots of big white eggs.'

Whipple's painting of Reginald Appleyard's winning pair of ducks.

Breeds of Duck

The Silver Appleyard – a breed recently refound.

Amongst the many breeds we keep at Folly Farm, the Appleyard is the first to start laying in the spring. The drakes are quick to mature and make excellent table birds with a distinct and most palatable flavour. The large Appleyard is a good example of the dual purpose bird, bred for both meat and egg production. They are quiet and one of the most decorative of domestic breeds. I am unable to find a reason why, in view of all these qualities, the Appleyard surprisingly lost its popularity. Today, we are not only indebted to Reginald Appleyard's achievement and skill as a breeder, but are delighted to make some contribution at Folly Farm towards the perpetuation of this colourful, attractive, useful and friendly breed of duck.

The Buff Orpington (opposite).

Orpington

This is an ideal bird to breed as a hobby. It is an excellent egg producer and any surplus drakes make quite good table birds. This is important because there will always be a few birds in the Orpington breeds which will fall short of the Standard for colour. Drakes with light heads or ducks too light in colour are no good for showing. Buff is the most popular colour and the bird should be this shade throughout although, frankly, it is not an easy colour to perfect, which gives an added challenge to the breeder who wishes to show his stock. In the past there have been many colours of this breed, such as the white, black, chocolate and blue, which seem to have disappeared, but as this is a useful and popular duck I feel sure that these other colours will soon be revived. If you have a pen of Buff Orpington, remember that even if the male has a light head, he will still throw male progeny with the desired dark, seal-brown head, which is the correct Standard. One drake is sufficient to run with five ducks. It was Mr W. Cook of Kent who bred the first Buff Orpington which he intended to be a dual purpose breed.

Magpie

The Magpie is very striking in colour with its bold markings. It is of medium size, beautiful, hardy and useful. The colours can be black and white, blue and white or even dun and white, but here again, to breed this bird with perfect markings is not easy. To achieve near perfection, however, will give the breeder great satisfaction.

The head and neck should be white, surmounted by a black, blue or dun cap covering the whole of the crown of the head to the top of the eyes. The breast should be white and the back should be covered with a heart-shaped mantle the same colour as the cap. Both the wing primaries and secondaries should be white. It is very difficult to breed these markings with good definition. One or two ducks in a brood may be reasonably marked but splashed down the neck and possibly across the breast. These will still, of course, be attractive ducks though not suitable for exhibition. It is easy to be left with too many mis-marked ducks and the best way to reduce the number is to sell them merely as ordinary, prettily marked ducks, never giving the impression that they are Magpies, which would mislead new breeders in understanding the correct breed Standard.

Magpies – noted for their striking colour and bold markings.

The true exhibition type Pekin.

Pekin

Originally bred in China, the Pekin reached this country in 1874. There are very few good examples of the exhibition strain at the present time, though it appears to have grown from strength to strength in America where the breed is used in the commercial market as high class table ducklings. The same conditions as for the Rouens and Aylesbury apply if fertility problems are to be avoided. The Pekin will be best mated in trios. They are a very attractive bird, with an upright stance, broad round heads, giving a chubby appearance, and lemon-yellow plumage. I feel quite sure they will make a come-back in this country.

41

Muscovy Duck

Although referred to as a duck the Muscovy is much more like a goose in more ways than one. For instance, it is a grazer and eats grass in the same way as a goose, and unlike all other breeds of duck the Muscovy male has no curled feathers in his tail. Their laying habits are much the same as that of a goose and also the incubation period is longer, being thirty-six days. If the Muscovy drake is mated to other domesticated ducks the progeny are mules, birds which are sterile and which will not breed. However, the ducks are excellent sitters and cover a lot of eggs. I remember on one occasion when a duck had hatched her brood secretly, she came marching down the yard with twenty-one ducklings. Because of their determined sitting habits, they make ideal broodies for hatching eggs of other species. The drakes can become very large weighing up to 10-12lb (4.5-5-5kg), whereas the duck is quite small, around 5-6lb (2.25-2.75kg). They appear in many different colours – black and white, blue and white, pure white, black and bronze – but I like those with clearly defined markings similar to those of the Magpie. A good mating is one drake to eight ducks.

This duck, resulting from a cross between a Muscovy drake and a Swedish duck, is a 'mule', that is, she will be unable to reproduce.

The Muscovy duck, with similar markings to those of the Magpie.

The Muscovy drake.

4 Fencing

As the population of waterfowl at Folly Farm continues to increase, the present high price of fencing and housing has made it necessary to explore methods of reducing these costs. The solution we finally adopted may be of interest to anyone wishing to keep waterfowl for the first time, or to those replacing old worn out fencing.

In 1976, that very dry year, a large pool was created by damming the valley at Folly Farm with the clay that already existed there; subsequently a second pool was made. These were populated by domestic ducks and wildfowl respectively. In all, some 500 yards of fencing was required. It had to be as fox-proof as possible in a hunting county where a large population of foxes exists. Conventional chain-link fencing was not possible due to its high cost, so it was decided to use 6ft (1.8m) light wire netting with a 2in (5cm) mesh. This is supported by 8ft (2.4m) posts that are sunk 2ft (60cm) in the ground, leaving 6ft (1.8m) above ground level. The posts are spaced at a distance of 7yd (7m) from each other. There is a single strand of wire strong enough to support the netting at the top and a second strand close to the ground. At the bottom, the netting is folded outwards along the ground.

The key to the success of this cheap fencing in keeping out foxes is entirely due to the use of an electric fence extending right around the perimeter. It consists of four, light gauge, stranded wires spaced about 9in (23cm) above each other with the bottom wire 4in (10cm) from the ground. The wires are mounted on insulated stakes with adjustable holders. The electric fence is outside the wire netting fence at a distance of about 5in (13cm). A strip of ground 1ft (30cm) wide, directly underneath the fence, is periodically sprayed with a foliage killer to stop the vegetation growing up and causing the fence to short circuit. The Gallagher E8 electric fence unit has been found to be satisfactory. Instead, however, of relying on the usual earth return, the negative is connected directly to the wire netting, so providing an excellent conductor especially when the soil is dry. During the hard weather, fox

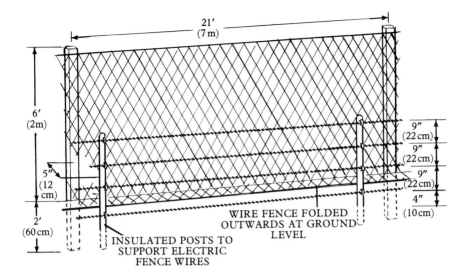

Fig 2 *Perimeter fencing with electric wires to deter predators.*

pad marks were found in the snow, but never closer than 2ft (60cm) from the fence. It seems that the foxes have been conditioned to keep well away from the electric fence and are becoming too nervous to approach it. Of course, the efficiency of this fence depends very much on the operator. At Folly Farm, the fence is checked every evening at dusk to make sure that all is well.

The other difficulty that has to be watched is snow, which can, especially with drifting, become a problem with any type of fence. When the depth of snow reaches the bottom wire it is disconnected, leaving only three wires in action. The snow seldom reaches the second wire but when it does, it usually only affects short lengths due to drifting. When this occurs, the new snow is shovelled away, exposing the second wire but this is never a big problem. If the fence is short circuited in this way, for a time the foxes are still too nervous to approach closely. Our fencing has now been in place for many years and so far no fox has managed to enter the compound.

For the breeding sections of the domestic fowl, each breed must be segregated. Each pen of birds has to be shut in at dusk and only let out when the eggs have been collected at about 9 a.m., to avoid theft or damage by predators such as rooks or crows. Because they have to be

A close-up of the external electric fence. This is the cheapest and most effective barrier against predators.

The fox-proof perimeter fence completely encompasses the holding lakes.

shut in at night there is no advantage in using an electric fence here. The breeding stock houses are extremely simple arks, as illustrated in Chapter 2, made of second-hand wood, without a wood floor but fixed to a mat of wire netting.

These simple and cheap means of keeping domestic waterfowl seem to be all that is needed. It is important that people should not be discouraged from keeping waterfowl by unnecessarily high costs of fencing and housing. It is hoped, therefore, that the methods described above will enable more people, who may have been deterred by the cost of ancillary requirements, to keep waterfowl.

5 Incubation and Rearing

Seeing ducklings hatch should be a joyful occasion, so I will explain the process of hatching and rearing ducklings in the hope that first-time breeders may avoid some of the disappointments that can occur with inexperience.

COLLECTING AND STORING EGGS

If you are collecting eggs for incubation, whether under a hen or in an incubator, the eggs should be gathered every morning and be kept as clean as possible. If they are a little soiled, just wipe them over with a damp cloth and finish them off with a dry one. Examine every egg for shape and colour so as to eliminate any egg with faults that may be perpetuated by future generations. At Folly Farm we mark each egg as they are picked up from the individual breeding pens with an abbreviation, such as W.R. for White Runner, B.R. for Black Runner or B.O. for Buff Orpington. For different strains they may be marked B.O.1 or B.O.2 accordingly. The eggs should be stored in a cool, even temperature and turned every day. As we hatch weekly at Folly Farm, all the eggs are in the incubator by the seventh day. You can keep them up to a fortnight before placing them in the machine, but I've found that the fertility does have a tendency to fall if kept longer than ten days.

NATURAL INCUBATION AND REARING

Those who have miniature Appleyards, Call ducks or Black East Indians will find that they sit very well and make good mothers. Some of the larger varieties, however, will sit but they are not sufficiently

The old-fashioned 'gamekeepers' coop and run.

reliable, in which case I believe that the broody hen is more suitable. I find that any broody hen will do but the Silky × Sumatra Game cross is a fine hen to use and is more likely to sit for the twenty-eight days required to hatch duck eggs. Chickens take twenty-one days.

The old 'gamekeepers' type of coop and run is ideal for the purpose of hatching eggs under a hen. Make sure that the coop has been well creosoted and disinfected long before you put your bird in. If it doesn't have a removable floor, it is best to put one in to avoid rats or foxes burrowing underneath.

First, cut a nice moist turf, 2-3in (5-8cm) deep, and put it in the coop. Beat it into shape so that the eggs will roll to the centre, then cover the turf with a little straw. When you put the hen in, it is a good idea to dust her with a louse powder because a bird will, naturally, never settle well if it is infested with lice. Put a few ordinary eggs in the nest so that the bird starts sitting well. When she has started brooding you can then remove those eggs and replace them with, what you hope are, fertile duck eggs.

Leave the hen for about twenty minutes, then go back and make sure she has settled down well on the eggs. If it is early in the season, there should be enough moisture in the turf to hatch the brood satisfactorily, but if it is later and the weather becomes hot and dry, it is a good idea to pour a little water in at the back of the coop to keep the turf moist.

At nine days you can test the eggs for fertility by placing each egg over a light. A torch will do. Those that are infertile will be clear and you must remove them, leaving only the fertile eggs for the hen to hatch. She should be allowed out every day, at the same time if possible, and be given a good feed of grain and plenty of fresh water. You can allow the

Eggs for incubation – a large Toulouse (top) and (left to right) a Miniature Silver Appleyard, an Indian Runner, a Cayuga, an Aylesbury.

bird a quarter of an hour but it is important to get her back on the nest within about twenty minutes so that the eggs do not get too cold.

Make sure that the run is covered to keep the ducklings dry when they venture outside and to provide shade from the sun. When feeding them for the first time, put down a shallow tin and scatter some chick crumbs on it. If you are feeding meal, it is important to wet it to a consistency that the ducklings can eat easily. I personally prefer to feed them dry chick crumbs which they seem to eat with no problem. The most important point to watch during the first few days is that the ducklings do not get over wet. If they are hatched under a hen, remember that they will have none of the natural oils which they seem to acquire from the feathers of the duck as she broods them, and which help to protect them from getting over wet. More ducklings are chilled and die through becoming damp from rain or even falling into a water dish that is too deep during their first day or two, than from any other hazard. I like to put some clean pebbles in the ducklings' shallow water container so that they can take the water from between the pebbles. This helps to stop the ducklings from getting over wet; I do this for several days.

As they grow you can use a fountain type drinker which it is worth placing on some wire mesh, an inch or two above the ground, so that the area around it doesn't get too messy. When the ducklings are a few days old, it is a good idea to remove the floor of the coop in the daytime and let the hen and ducklings stay on the grass. At night, replace the floor, together with a fresh sprinkling of straw. This will keep the ducklings dry, healthy and happy for the night. Make sure the coop front has plenty of ventilation holes. Ducklings hate to be cooped up too tightly and can soon get overheated.

At two or three weeks you can remove the hen and instead of using the run outside the coop, you can replace it with a roll of wire about 2ft (60cm) high. As the ducklings grow, extend the wire, moving them onto fresh ground every day until you take them from the coop at about five weeks old. Put them into their final, larger house, where they can have free access to water. At this stage, you can also change their food to growers pellets which are quite satisfactory until the ducklings begin to feather up properly when you can introduce some grain. Remember that all ducks need mixed grit which should be freely available. When they are fully feathered, ducklings do not require shelter for health reasons but will still need to be shut in a well-ventilated house at night,

Incubation and Rearing

A fountain drinker and shallow feeding dish.

to protect them from predators.

If you have a small Call duck, Black East Indian or Miniature Appleyard, then the time that they feather properly is when they begin to fly. I always clip a few flight feathers from one wing which stops them from taking to the air until the feathers grow again. If the birds change hands at this point and their new environment is correct and they are in pairs, then when the feathers grow again they will fly but will return to their new home. There is no need to clip the feathers of larger ducks because their maximum flight in a high wind would only be two to three feet, so quite low wire netting will keep them under control.

ARTIFICIAL INCUBATION AND REARING

If you have fertile duck eggs available and no duck or hen under which to brood them, then it is well worthwhile trying an incubator. There are

52

a number of small, suitable incubators on the market these days.

Over the years, much has been written about the technical side of artificially incubating duck eggs. I am inclined, however, to leave this subject to the manufacturers. I feel it is much more important to concentrate on the preparation of the eggs, for no machine, however good, can hatch badly stored eggs or those from improperly fed stock. The breeder should place in the incubator fresh, fertile eggs with good, sound shells, laid by mature stock that has been properly fed and cared for. So many beginners are ready to blame the machine for their failures when they should be looking at the management and preparation of their breeding flock.

Do follow the incubator manufacturer's instructions as closely as possible. Warm the eggs up slowly and make sure the machine is running at an even temperature 103-103.5°F, certainly not more than 104°F. Open the door once a day for cooling. The length of time that you leave the door open depends on the temperature of the room, but between five and ten minutes seems to be sufficient. When the door is open you can test for fertility and remove any bad eggs – the gases from these can be fatal to the good ones. There are many types of incubator but whichever one you use, the correct amount of humidity must be maintained at all times. Each machine is different and so is the room in which it is placed, so first time success cannot be assured. All machines will hatch eggs but they are only as good as their operators, and only through practical experience with your machine will you gain success.

If you have success with your hatch then the next stage is to rear them on. I find that ducklings rear very well on a wire floor and we use an old-fashioned tiered brooder. Another method is to put the ducklings under heat lamps, this being the cheapest way to start. When you place them under any form of heat, the ducklings will soon make it clear whether the temperature is correct. They will crowd under the source of heat if they are too cold and will move away if they are too hot, so adjust the temperature accordingly. At about ten days to a fortnight they can be moved to a hay box, which is another old-fashioned but proven method. The hay box consists of a small ark with a run attached. A light sack of hay is placed on a frame inside the house about 10in (25cm) above the ducklings to regulate the temperature, and a wire floor fixed underneath. Depending on the temperature, you remove hay from the sack over a period of a week to ten days. By this time, they should have cooled

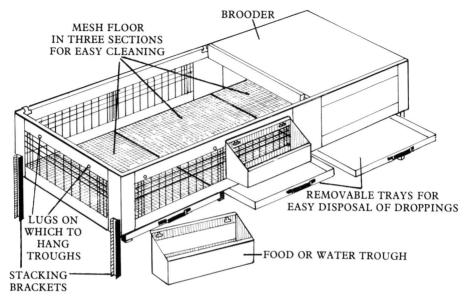

Fig 3 The Eltex tiered brooder.

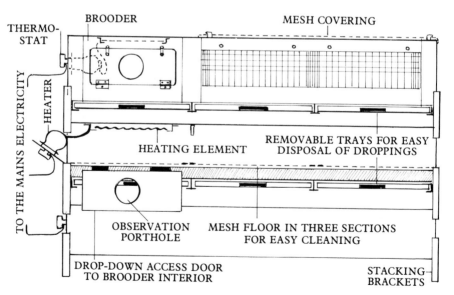

Fig 4 Cross-section of the Eltex tiered brooder.

54

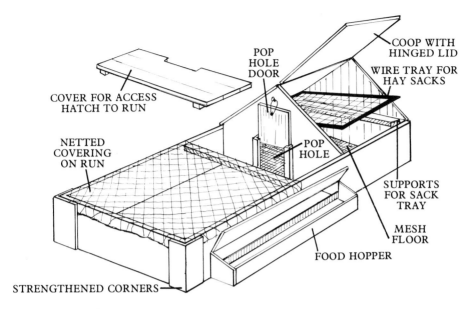

Fig 5 A hay box.

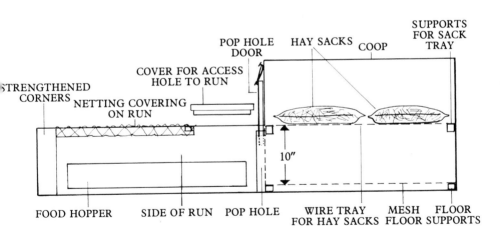

Fig 6 Cross-section of a hay box.

out and can be moved to their final quarters at about one month to five weeks old.

Recently, it has been the trend for parents to encourage the younger members of the family to take up keeping ducks as a hobby. This is to be applauded, providing that parents keep a watchful eye on the procedure, and I wish them a fair amount of beginner's luck because as anyone who keeps livestock knows, there are disappointments. This should not deter but act as a spur to improving your methods, and help to emphasise the joys of success. The experience should help to teach youngsters a respect for nature's ways.

6 Keeping Domestic Geese

Geese are certainly one of the oldest domesticated fowl and have been kept in all parts of the world for centuries. They are extremely intelligent, but their greatest asset is that they are grazers. Geese are becoming more popular as pets, being generally companionable creatures. In addition, they are seldom troubled by illness and have a very high resistance to disease. Some breeds are quite beautiful and can be kept as ornamental birds on ponds and lakes. You may even wish to breed top quality birds for exhibition, which has the added interest of competing at shows with the possible satisfaction of winning a show championship. Here again we will concentrate on keeping on a small scale as a hobby or for the home market.

ENVIRONMENT

Space

Consider carefully the area you have available for keeping geese. Remember that 80 per cent of their food can come from grass, so they need considerable grazing land. As a rough guide, a quarter of an acre will support five large adult geese, depending on the location and, therefore, the quality of the pasture. Remember also that some breeds are better foragers than others.

If you hope to let your geese hatch their young, think about how much space you have before deciding to rear them, remembering that there is always a market for day old goslings. Overstocking a piece of ground creates the worst possible conditions, so do not try to rear them if you are pressed for space.

Housing and Water

Geese need shade in hot weather and, if the area is bleak, benefit greatly from being supplied with some sort of shelter such as wattle hurdles or straw bales. Their housing need only be very simple, with a floor area of about 7ft (2m) square and a height of 3-4ft (1m) for a trio of large geese. Make sure there is plenty of ventilation. At Folly Farm we make the ark doors of a mesh big enough to allow plenty of air but small enough to prevent the geese pushing their heads through. Of course, stables and outbuildings of any kind can make quite satisfactory housing.

One of the most important reasons for housing geese is to protect them from predators, especially foxes. Foxes will be likely visitors, so care must be taken to shut your geese up, well before dusk and not let them out too early in the morning.

Swimming water is only absolutely essential for the large varieties of geese such as the Embden and Toulouse. This provision encourages and assists these heavy geese to mate.

SUCCESSFUL BREEDING

Whichever breed you choose to make up your pen, the first year of breeding is not always successful as often the eggs do not hatch out well. Don't be disappointed by this or jump to the conclusion that there is something wrong with your geese. It is not until the second year that you can give a true judgement of their performance, either as producers of fertile eggs or as good natural parents. Remember, if they perform well in their second year you may have ten, twelve or more years of continued success.

At Folly Farm we like the geese paired as early as possible, certainly well before Christmas. It is usually easy enough to introduce a gander to his geese, but be sure to do this with them all at the same time. Trying to introduce a goose at a later date is very difficult. The original set will usually ignore the newcomer and even chase it away – most likely your attempt will end in failure.

If you decide to choose a breed to sit and hatch their own eggs, I think it is wise to go for the medium or lighter breeds. The heavy breeds tend to be far too clumsy, often breaking their eggs or treading on goslings

Geese look very attractive on water, especially the Chinese.

when they hatch. So with the heavy geese it is certainly worthwhile considering putting the eggs under hens. The method of storing goose eggs and the preparation for putting them under hens is identical to that described in Chapter 5 for ducks. You will be able to place four or five eggs under a hen. Of course this depends on the size of the eggs and the hen. With goose eggs, however, when the hen is liberated each day for its regular feed, the eggs can be examined and turned by hand, as they are obviously too large and heavy for the hen to perform this task by herself.

Incubation

If, as so often happens early in the year, you have some eggs you wish to set and no broody hen available, it may be worthwhile for you to try an incubator. I have found no difficulty hatching the lighter breeds of goose with the Mayfair incubator that I have used for many years. I must admit, however, that hatching the larger breeds, such as the Embden and Toulouse and especially the extra heavy, top quality birds, I have found rather disappointing. The last three years have shown a steady increase in the number of goslings found dead in shell during the last week of hatching and I have spoken to other breeders with different machines and methods of hatching who have also been stricken with the same recent rise in the incidence of this problem. On the whole, reports I have heard seem to give me the impression that better results are achieved by putting the eggs under a hen than by using an incubator. Although we are continuing to try and find the answer to this problem, so far our endeavours have proved unsuccessful. Lighter breeds of geese will take up to twenty-eight days to hatch, the heavier breeds taking up to thirty days and even a little longer.

Feeding

Use any good proprietary brand of chick crumbs for feeding goslings, making sure they have no added medication. I feed ad lib, up to the point of feathering, and from then on I feed layers pellets, as much as they can take in the morning in about twenty minutes, scattered on a clean part of the run and the same process with grain at night. As with keeping ducks, make sure they always have mixed grit freely available.

Rearing

When the goslings hatch, again you can follow exactly the same procedure as for ducks, although unlike ducklings, they benefit considerably from being placed on a lawn where there is plenty of short, crisp, green grass. It is not many days before they start cropping away at it. Whether you have ducklings or goslings, always keep the coops very close at hand so that you can keep a watchful eye on them at all times of the day. Make sure they have shade when it is hot, and slip out to cover the runs if there is a thunderstorm. Take care that no goslings fall on their backs, as at this early age they find it almost impossible to upright themselves and, if not assisted, would very soon die.

At this stage, goslings grow very fast indeed and eagerly crop grass. If they are in a coop and run they should be moved onto fresh ground twice a day. As they grow in size, the coop will obviously become too small and it is very unwise to shut them up in the house at night with insufficient air. So at this point you can dispense with the hen and put the goslings into a larger house, with adequate clean litter and plenty of ventilation, still letting them out daily and shutting them up securely at night. They will soon feather up and mature.

7 Breeds of Geese

LIGHT BREEDS

Chinese Geese

These geese came from an area around China, Siberia and India. They have been developed over the centuries from the Swan goose. There are two Standard colours: pure white with orange knob and bill and the fawn variety with similar markings to the Swan goose. They normally lay some sixty eggs a year but have been known to lay as many as 100 in exceptional circumstances. Of all breeds, the Chinese are wonderful watchdogs. In fact, at Ballentine's whisky distillery, near Dumbarton in Scotland, they use this breed to guard their stocks of valuable, maturing whisky! They give warning the moment a stranger comes near and will keep calling.

They are intelligent, friendly with their own species, and make excellent mothers and very proud parents. The goose is a good sitter and, allowed to carry on naturally, she does a fine job of rearing her brood. The gander also makes a protective father, always keeping a watchful eye on his family, ready to guard them from danger. The eggs are quite large for the size of the bird, mild in flavour and excellent for cooking. Besides laying as normal in spring, a number of eggs may well be laid in late September or October.

Although water is not essential for their well-being or for mating, these geese spend much time diving and splashing about if it is available. As probably the most ornamental of all the domestic varieties, it is pleasing to see it swimming on the water, where it can display its true beauty.

I run two breeding lines of Chinese geese at Folly Farm. The Old English are generally more stocky and less colourful than the line introduced by Christopher Marler, which he imported from America. The American strain has greater ornamental characteristics and the

This shows the inherited swan-like characteristics of the fawn Chinese.

cross between the two is producing considerable hybrid vigour; it is now beginning to appear at the shows. The Chinese goose is very hardy and does not need a lot of shelter but its eggs require protection from predators such as crows and magpies. Running more than one gander in a flock will give mutual confidence when they act as lookouts, and between them they will raise the alarm by making much more noise than when running singly. A gander may run with up to as many as five geese.

In addition to their beauty, Chinese geese are always quick to raise the alarm at the approach of strangers.

MEDIUM BREEDS

Roman

This variety is well worth considering for those who do not have a great deal of room for grazing, as this should be the smallest goose. It has pure white plumage, a delightful outline and orangey-pink bill and legs. They are a very active breed but can be a little noisy as they are apt to chatter a lot – so consider your neighbour before you make this choice. Romans are ideal for those who wish to let them sit and hatch their own eggs. They do not require water to assist in mating being such a light breed, and are not so likely to break their eggs or tread on their

The Roman is the smallest breed of goose, ideal for those with limited space.

young when they are hatched. They make excellent mothers and are a pleasure to see walking around tending their family. It is important that this goose be kept small. The gander should certainly not weigh more than 14lb (6.35kg) and the goose not more than 12lb (5.5kg).

Buff Geese

AMERICAN BUFF

The American Buff goose originated, of course, in America. It is a beautiful buff colour with deep orange beak and legs, showing its best during the autumn at the time of the shows. Unfortunately, this colour

The American Buff is larger than the Brecon Buff and has orange beak and legs.

has a tendency to fade in the summer sunlight, so it will lose some of its beauty progressively through the sunnier months unless kept in the shade.

BRECON BUFF

The American Buff should on no account be associated with the other buff goose, the Brecon Buff, which is quite a different breed. This was originally bred by Sir Rhys Llewellyn in 1928 from some buff geese found in a Breconshire hill farm. The Brecon Buff has a deep pink beak and legs and is a lighter and more refined specimen with a slightly pinkish tinge to the buff colouring. I find visitors are very attracted to this goose, partly because of its beauty and partly because of its docile and friendly nature. Both of these geese are good layers and one gander will run quite successfully with three geese.

The Brecon Buff has pink beak and legs.

Sebastopol

These geese are not very common in this country so if you wish to breed something really unusual then these would be a good breed to try. They are a medium-sized goose with long, curly feathers. Those who have not seen this breed of goose before often ask whether the goose is moulting. When it is explained that the bird is naturally feathered like this then people usually begin to appreciate its beauty. There are two strains of this goose. One has the curly feathers right up its neck and the other has a plain neck but the feathers flowing from the back are longer and, to some people, represent greater beauty. Both of these types are acceptable for exhibition.

A low fence will keep them under control as they are quite unable to fly, but as their feathers trail the ground, it is necessary for them to have swimming water to keep clean. Anyone wishing to take up the breeding

67

The Sebastopol, commonly referred to by children as the 'Pantomime Goose', is sometimes coloured, but the preference in this country is for pure white.

The long curled feathers of the Sebastopol goose.

of this quaint and unusual goose will, I'm sure, be heartened to know that we have found them very easy to keep, hardy, a free breeder and very fertile.

West of England

The West of England goose, together with its descendant the Pilgrim goose which travelled to America with the Pilgrim Fathers, are truly unique, because they are the only breeds that can be sexed at the time of hatching. The females of these breeds carry distinct grey markings and the males are completely white. This 'auto-sexing' can be most helpful to beginners who have yet to learn how to physically sex geese.

As the ganders are always white and the females have a grey saddle, breeders can easily identify the sexes of West of England goslings at day old.

HEAVY BREEDS

Dewlap Toulouse

This goose originates from the Toulouse region of France and was first brought to Britain by Lord Derby in 1840. In France, the emphasis has been on the production of fat livers for pâté de foie gras and has developed along very different lines to the British breed. It is an imposing, mammoth breed of goose with massive features and a stately carriage. The pure bred type should not be confused with the common grey goose which is sometimes called Toulouse because of its grey plumage but which is usually a cross-breed and bears no resemblance to the pure Toulouse breed. Owing to its great size and unusual features, the Toulouse has remained the aristocrat of domestic geese. Few breeders of waterfowl have not desired to own a pair of the handsome Toulouse.

The wild Greylag – ancestor of the modern Toulouse.

The true Toulouse has a blocky, rectangular body mounted on short legs, a full paunch, deep keel and a large dewlap. Plenty of keel and gullet are the desirable features and it is not easy to breed these to include as much width and depth of body as is possible without losing fertility for a first class exhibition specimen. The straight, deep front was developed to a greater extent in the American strains. Birds imported from that country by Christopher Marler helped to improve the conformation of our British stock.

Water is essential for the Toulouse to be able to mate. If you do not have a natural pond or stream it will be necessary to build a bathing pool about a foot deep. This need not be large but it will certainly help you to achieve the desired results. We like to run our geese in pairs but one gander to two geese should be quite satisfactory. Give them plenty of room for exercise and don't let them get over fat. Owing to their size, weight and awkwardness they do not make good mothers; moreover, a Toulouse is almost a non-broody. A strain of sitting hen that will go

Despite their massive size, the Toulouse have a quiet and gentle nature.

broody early in the year will be much more effective.

The photograph is of our six year old Toulouse who is an excellent example of the breed and over the years has become a star! She has a gentle manner and never fails to capture the attention of the thousands of visitors who come to Folly Farm. She is a great favourite with the media and has now appeared in no less than eleven publications, including such nationals as *The Observer*. She was photographed in full colour for the front cover of the banker's magazine *Coutts and Company* and, of course, with such beauty she was bound to appear in *Vogue*! She has also appeared eight times on television. No goose could do more in an exercise of public relations for the world of waterfowl!

Embden

As a breed, the Embden is extremely hardy and a good forager. The goslings will mature fast, are easy to rear and, once they are feathered, will not require housing. However, unless you have fox-proof fencing, as described in Chapter 4, they will have to be housed at night to protect them from foxes. I have heard it said that a well-grown Embden gander will protect his wives and soon chase off the intruder but I would certainly not take that risk from my experience of the foxes in this area!

When the goslings are first hatched a beginner may expect all of them to be yellow as is customary with white birds. You may be surprised to see that some of them have grey down on their backs, but don't worry – the grey colour will disappear as the bird grows its adult plumage. Some will hold a few grey feathers on the rump and this often denotes a female gosling.

The Embden is an outstandingly beautiful bird with its snow white colour and large, yet compact body. The bird has a perfectly clean outline, free from any signs of keel or gullet – the features of the Toulouse but very bad faults in an exhibition Embden. If there are any signs of these it probably means that the bird has, at some time in the past, been crossed with a Toulouse. This fault must be avoided at all costs if you wish to breed winning stock. Like all

Our six year old (opposite) is an excellent example of the Dewlap Toulouse.

The Embden is the most widely used commercial breed, but top quality exhibition stock is hard to come by.

white waterfowl, they will never look their best unless they have plenty of water in which to swim and wash. Also, as with the Toulouse, the water helps to encourage successful mating. Three or four geese to one gander should prove a satisfactory breeding pen.

Buff Back and Grey Back

The Grey Back is a medium sized, stocky goose with very attractive markings. The grey colour on its head should extend well down the neck and it should have a grey mantle spread over the back, with grey feathers on the thighs. The colour should be well defined and evenly distributed on a snow white ground. The general markings are the same for the Buff Back but with beautiful, varied shades of buff. These strikingly beautiful birds were quickly added to our collection at Folly Farm contributing enormously to the interest shown in the variations of goose types and colours.

The attractive markings of the Grey Back (above) and the Buff Back (below) add considerably to the interest of our collection.

8 Handling, Sexing and Sickness

HANDLING

Ducks are usually quite easy to handle once they are caught, but catching a single bird from a pen of layers must be accomplished with as little disturbance as possible. I have found that the most effective method is to single out the bird you require, quietly drive it to a corner, and then catch the bird using a fisherman's landing net with a 2 ft (60cm) diameter ring, attached to a 4 ft (1.2m) handle. As the bird runs down the side of the fence, place the net quickly in front of it and your duck will run into the opening. I have found that this procedure does not panic the bird and is least upsetting to the rest of the birds in the pen. Unnecessary chasing round the pen can not only put them off lay but make them very difficult to catch on future occasions.

When retrieving the duck, take it firmly but gently by the neck and withdraw it from the net. Never take hold of it by its legs or wings. Draw the duck towards you and rest it firmly on your forearm, leaning its body towards your chest and laying its head outwards over your elbow; slip one leg between your thumb and forefinger and the other between your other fingers and lay your free hand lightly across the wings. In this manner the duck will lay quietly on your forearm and experience the least stress.

Geese can be easily driven into a corner of their pen. From here, to catch a goose, position yourself so that you can grasp it firmly by the neck with your left hand (at which point the goose will usually squat), leaving your right hand free to reach over its back, your arm securing the wings. Lift the goose and transfer your left hand from its neck as you tuck that under your arm and grasp the legs. Now your right hand can be held across the goose's back and wings.

A duck should be held with its body towards your chest, laying its head outwards over your elbow.

A goose should be held with your left hand grasping its feet and your right hand positioned across its back and wings.

77

IDENTIFICATION

For those of you who have chosen to take up the breeding of pure strains, a form of identification is essential. You must be able to record the various lines you will be using in the progressive breeding programme. Wing tags can be applied which remain clean and are easily read. The tags should be marked with the initials of the breeder on one side and the duck's individual number on the other. The tag is positioned at the front of the wing and is covered by feathers so that it is not unsightly and can easily be exposed to be read by parting the feathers. It is also a great asset to have permanent identification when taking your ducks to shows, for it is conclusive evidence that the duck is yours, should it escape or be put in the wrong pen. So far, none of my birds have been hooked up and damaged as I have seen happen with leg rings. In my opinion wing tagging is the simplest, safest and best way of identification of waterfowl.

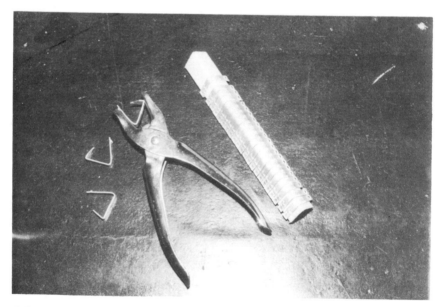

Wing tags are the best form of identification for waterfowl.

SEXING

At Folly Farm, we see no point in sexing day old ducklings as they may be sold at a later date when we know whether they are up to our standard. Over a period, the numbers of male and female are about equal, and as they are sold in pairs there is little to be gained by sexing them at day old. As the ducks become fully feathered, catch one and hold it for a minute or two. It will soon announce its sex with either a deep sounding quack, which indicates that it is female, or a high pitched hissing sound, indicating the male. Also, the drake has a few curled tail feathers at the top of the tail.

The sexing of geese has always been a problem. You may well arrive at some indication of sex by observing your geese at particular growing stages, such as the general masculine appearance of a gander, its slightly larger head size or its tendency to hiss at a passing dog. These are good indications, but indications *only*.

A goose held breast uppermost for sexing.

Before you can be even ninety per cent sure of the sex of a goose, the bird must be fully matured. Then, take the goose, turn it breast uppermost and place it between your knees, leaving both hands free to support the tail and turn the feathers and fluff away from the bird's genital organs. Allow your thumbs to gently prise open the vent with a gentle but firm downwards and outwards pressure. Under ordinary circumstances, with matured birds, you should be able to expose the male organ. This is not always so; even the most experienced find it difficult to sex certain geese. If this is your first attempt at examination, on no account must you use force, every movement must be careful and gentle – in this way you are much more likely to achieve success. There is only one way to learn – through keen observation and plenty of practice.

SICKNESS

Waterfowl are subject to very few diseases and at Folly Farm we always assume the policy of prevention rather than cure. If a bird becomes off colour within our flock we dispose of it immediately – an infrequent occurrence, I'm glad to say! We never try to cure a bird (unless we are convinced of the reasons for the bird's ailment) or struggle to rear poor or second rate ducklings, for these may find their way into the breeding pen where there is every chance they will pass on their particular weakness to their future progeny. In this way we are only copying nature, leaving the fittest to survive and perpetuate the breed. By making sure you have purchased your birds from a breeder with good healthy stock, with his assurance that they are not too closely related, and ensuring that you apply good management, the stock should have a high resistance to disease. If illness should occur, isolate the sick bird and call in professional advice immediately.

From time to time it is absolutely necessary to kill a sick duck. This can be carried out quickly and humanely by taking the bird's legs in your left hand and taking up its head in your right, with your fingers under the bill. Bend the head back and with a positive downwards thrust you will find that the head will easily dislocate from the neck.

Owing to their weight and length, killing a goose is slightly more difficult. In my opinion, the quickest and most humane method is to take the bird by the legs, with its head pointing away from you, and place a broom handle across the back of its head. Putting one foot on the broom handle, take a leg in each hand and place your other foot on the handle *at the same time* as giving the goose's legs a sharp upward pull. This will dislocate the head from the neck leaving a gap into which the blood can drain. This is not an easy manoeuvre, so additional help may be an advantage.

9 Showing Waterfowl

In the period following the War when interest in waterfowl was at its lowest, very few waterfowl appeared at the poultry shows. They were certainly not taken seriously by the majority of the showing fraternity. Waterfowl, and especially the Runners, were usually placed right on the floor under the chicken pens, where they were most difficult to see and almost impossible to judge. It was not until the seventies that interest in breeding and showing waterfowl really started to grow. Also, by that time, the British Waterfowl Association had re-established itself into a strong and active organisation and was capable of making its voice heard whenever the opportunity arose to promote the breeding and showing of domestic waterfowl.

Realising the necessity to import new stock to help invigorate our dwindling British bloodlines, Christopher Marler, a leading breeder from Buckinghamshire, bought in waterfowl from America where they had continued to upgrade their stock at a time when a handful of British breeders were struggling to maintain our breeds, let alone improve them. This importation did much more than merely improve the stock with new blood, it revived the competitiveness of established breeders and captured the interest of newcomers. Since then, the showing of waterfowl has grown from strength to strength. Today it has reached an all time high and there is no better way for a beginner to start than to join the ranks of the British Waterfowl Association, which will send you their journal containing the dates and venues of their regional shows. These occur every autumn when the birds are in perfect colour.

Purchase a British Waterfowl Association's *British Waterfowl Standards* book and make a careful study of the birds you wish to keep. It is a good idea at this stage to become a steward to one of the judges at some local shows. You will find that most judges are helpful to their steward, probably discussing some of the points as they move from bird to bird, and a good deal can be learned in this way. There is nothing quite like the visual impression of the live bird to imprint on your mind the

Some of the top awards won by Folly Farm waterfowl in the 1985 season.

required winning characteristics for which you are looking. If you have decided to show your waterfowl you will, no doubt, have purchased your stock from a reputable breeder with award-winning strains. You then must select the bird you think is most suited to being a possible winner, making sure that it is in top condition and well feathered. Obtain a copy of the schedule and entry form from the show secretary and enter your bird.

CLEANING

Presentation of your bird is extremely important and preparation should start weeks before. First keep the home pen clean and free of mud. Avoid washing the feathers if possible as your duck, given access to clean water, will achieve a far better result by being allowed to preen itself. If washing is going to be unavoidable, then do not use any bleach

Showing Waterfowl

A converted tea chest is ideally suited for transporting ducks and geese.

or detergent as these will strip the feathers of their oils and remove the natural bloom which will take some time to restore. The way I prepare my show birds is to pick them up as cleanly as possible the day before the show and place them in an indoor pen on a deep litter of dry straw. This will ensure that when you box your birds the following morning, they are completely dry. Never box damp birds, this is a certain way for them to lose their natural bloom. Make sure the box has plenty of ventilation and don't overcrowd your birds. There is nothing worse than trying to clean up a mucky specimen upon your arrival at the show.

THE SHOW PEN

Most breeds do not need a lot of training for the show pen if they are treated quietly and hand fed. The exception to this may be the Runner, which tends to be of a slightly more nervous disposition. It is very important to familiarise them with the show pen well beforehand. A Runner, outstanding in his own surroundings, will often crouch nervously in a show pen. The best way to overcome this is to improvise a show pen at home and situate it in a place that you will pass by frequently during the course of the day. Install your Runner and each time you pass, throw it a handful of wheat. The bird will soon associate an approaching figure (whether it be you passing its pen at home or the judge at a show) with food and stand up to its full height in expectation, presenting itself most impressively. Thankfully, these days most show organisers pen their Runner classes properly, but at one time it was not uncommon to find this breed at the bottom of a stack of show pens and, what was worse, with no backs to the pens! No Runner, however well trained, will stand up in these conditions. It was very disappointing for the exhibitor with a good bird who felt he had been wasting his time.

At some of the larger shows, you may be allowed to pen your birds the night before. In this case, make sure they have enough food and water for the night. Grain or pellets are suitable but avoid meal or mash as these will stick to the bill and feathers and ruin the presentation. On the morning of the show, feed little or no food, in particular to your Runners, as a crop full of food will distort the shape of the bird and considerably reduce your chances of success.

The basic requirements for the show pen apply equally to duck and geese. The latter on the whole, adapt more easily to the showing environment with the possible exception of the Toulouse. Some Toulouse will behave impeccably, showing themselves off beautifully on their home ground, but taken to a show will sulk and refuse to stand. They bow their heads, looking quite sick, which can be embarrassing to the owner and, of course, lose them any chance in the eyes of the judge. A good goose that develops this behaviour can seldom be cured and will be a great disappointment to the breeder. Some duck and especially geese will have the ability to present themselves well in the show pen. They seem to have a charisma that is so hard to define but which never fails to attract the attention of the judge.

Once you have prepared your bird to the very best of your ability and have placed it in the show pen, you must keep well away until the completion of the judging. You may return to the pen afterwards to see if your bird has been placed and, whether or not you have received an award, now you have a wonderful opportunity to compare your bird

Breeding pairs in their show pens.

Geese penned for showing.

Owing to their size, these small duck are easier to handle and pen than the larger breeds, which is reflected in the popularity of these classes.

with others in the class. If you do not follow the reasoning behind the placings of the stock then ask the judge if he would explain it to you. Most judges are accommodating about this issue and should be only too pleased to help you learn the art of showing and judging waterfowl.

10 Management Round the Seasons

As a guide to newcomers, here we give an outline to the year's work involved in keeping ducks and geese, as carried out at Folly Farm. We hope the month by month format will assist first time breeders and help them to realise what management of stock entails.

January is as good a month as any with which to start, for it is a time when you can expect the first eggs of the season. Thoughts may occupy your mind on how you can improve the results of the previous year. Even after years of experience you find yourself eagerly awaiting the challenge of the new season. As every season is different, each year will need a slightly different approach. Of course, you will never be sure of the outcome, even if you have a good hatch, for it is not until the birds are fully mature that you will know if you have made good progress with your breeding combinations, or whether you are slipping back. Whatever the result, the following season will find you with a little more experience which will help you to improve your breeding programme. There is one thing certain when you become involved in the breeding of livestock, life will never be boring!

JANUARY

This can be a difficult month for livestock, especially waterfowl, but although Folly Farm is well over 600 feet above sea level and can be very windswept, I have never lost a young, healthy bird from the effects of severe winter conditions. This is mainly due to our very careful attention to ensuring that every bird has access to unfrozen water and is fed regularly. A plastic drum, split in half, is an ideal substitute to the natural but frozen water supply, as you can tap its sides with a hammer, if frozen, and release any ice that may have formed in the bottom before you refill it. This we do twice a day, if necessary, at feeding times. We feed layers pellets in the mornings, as much as the bird can consume in a quarter of an hour. By watching them, you can soon tell when their crops are full. Just before dusk, we repeat this feeding routine with wheat.

If weather conditions are very severe, we scatter a little straw on the ground just outside the house daily, for the ducks to stand on. Ducks lose most of their body heat through their legs and you will find that they will benefit greatly from having a little insulation against the frozen ground. As in this month our breeding stock begin to lay, in wet, cold conditions we also litter the houses down with some fresh, dry straw every day, cleaning the house out about twice a week.

Ducks usually lay at dawn and most have laid before 9 o'clock, so when we let them out, we are able to pick up the eggs at the same time and prevent them from becoming frosted, remembering to mark which eggs come from which pen. The eggs are usually alright when retrieved from the dry straw of the house, but we are careful not to get them

All types of duck live harmoniously on the holding lake.

frosted when they are placed in the bucket. We use a plastic bucket which is softer and warmer than a metal one, and get them back to the hatchery as quickly as possible for storage in a temperature of around 50°F. We use plastic egg cartons to store the eggs, placing them on a table that has been propped up at one end with a 4in (10cm) block. Every day we turn the cartons round so that the opposite side of the egg is now uppermost. This helps to stop the embryo sticking to the shell. Although at Folly Farm we like to hatch our ducklings as early as possible in the season, we do not usually start the incubators until February.

At Folly Farm, we have a large compound and lake on which we always hold a surplus of breeding stock that can be called upon at any time, if we should lose a bird in the breeding pens. Swelling these numbers are any birds that were not sold in the autumn and are kept on to be sold in the spring. These ducks are out all the year round in our compound without any housing whatsoever, and if the crows or magpies have an egg or two, we are not unduly bothered; the ducks are a mixed flock and the offspring would be cross-bred anyway. However, it is very important that they should have as much attention as the breeding birds. They are fed following the same programme as the penned stock, and every morning and night the ice is broken at the edge of the pool so that the birds may have a drink and a swim. I repeat that the ice should only be broken at the edge, to make access to and from the water easy. Once we know how much the flock requires, we feed them in this shallow water where they quickly dabble and pick up every grain, thus avoiding feeding a host of crows and jackdaws from miles around, who are quick to spot an easy meal.

One would expect the routine for breeders to be well under way before the end of the month.

FEBRUARY

No matter how cold the weather, ducks do not like to stay in their house, they like to be out in the air, although they do require shelter from high winds. A strong cold wind is the element that ducks dislike most, so if the area in which you keep your ducks has no natural shelter, put some straw bales or a sheet of tin against the fence on the side from which

Ice at the edge of the pool should be broken to allow easy access to and from the water.

come the coldest winds. At this time of year a biting wind can severely disrupt your breeding programme, no matter how good your management, by putting the ducks off lay and affecting fertility.

The feeding and watering routine continues as in January and, at Folly Farm, we have the additional chore of running the incubator, as we wish to hatch as many ducklings as early in the year as possible.

We open the incubators for about a quarter of an hour every day for cooling and on the seventh day, *after* this cooling time, we place the new, marked eggs in the machine. In the incubator room we have a regular and disciplined routine. First, we turn the eggs in store for the incubator, then we check the temperature and make sure that the thermostat is working correctly. We open the doors for cooling the eggs and top up the water trays. The correct humidity for your particular machine can only be arrived at after some practical experience. We have found through trial and error with our incubator that if the water trays are kept full during February this provides sufficient humidity to

*A Mayfair incubator, roomy enough to allow a tray
for each individual breed.*

achieve good results.

The Mayfair incubator that we use is a large machine costing very little more to run than a small one, giving us plenty of tray space and allowing us to set the eggs in groups from the various pens. This gives us easy, up-to-date information on how the individual breeds are progressing. St Valentine's Day or about the middle of the month is when our Chinese Geese also begin to lay. As their eggs hatch well in the incubator under the same conditions as the duck eggs, we can include them at the same time, on a separate tray.

MARCH

Each tray is marked with the date of the expected hatch, which should occur during this month. Two days before this date we transfer the eggs to the bottom of the machine for hatching, and as there is room to group them separately, we are able to distinguish the individual breeds. This is

An Eltex tiered brooder can be used for brooding artificially hatched ducklings for the first ten days.

most helpful with breeds such as the crested duck, because a percentage of these are hatched with no crest at all, so if they were hatched in a tray of mixed breeds, we would be unable to distinguish them from other breeds of the same type and similar colouring. Also, on the hatching tray is a very good time to select the best of the Swedish and Magpie duck, for it is not too difficult at day-old to pull out those ducklings that are improperly marked. There is no future for these in a well-organised breeding programme, designed to improve the breeds.

It is important not to remove the ducklings too quickly from the incubator; we leave them until the down is well fluffed up and dry. Over the years at Folly Farm, we have tried many ways of rearing ducklings and we have settled for what we consider to be by far the easiest and most successful method. We use an Eltex tiered brooder or one of similar design. The ducklings are placed on wire and warmed with a thermostatically controlled electric heat plate above their backs. The ducklings soon 'tell' us if the temperature is right. If they are drawn together in a heap beneath the plate then they are too cold and if the temperature is too hot then they retreat away from the heat into the run: they should be well scattered. It is far better to have them too hot than too cold, as there are too many ducks lost each year from being chilled. We have found that rearing them on wire avoids the problem of them getting too wet. We never like to see ducklings overcrowded and up to thirty in a group seems quite satisfactory. I put a small pan of water close to the heater for the first few days, just to start them drinking, with a sprinkling of chick crumbs on an egg tray to encourage them to feed. We do not like using fine mash as the ducklings have great difficulty in swallowing it – they manage the chick crumbs easily and soon fill their crops and maintain a very fast growth. After about two days, chick crumbs are fed in the feeders on one side of the brooder and water is put into the troughs on the other. They never seem to have any difficulty in finding either. We leave the temperature well up for the first five days before starting to reduce the heat. The droppings fall into the trays below and need to be cleaned out once a week. After the first week in the brooder the ducklings can be cooled down quite rapidly, and it is not difficult to see from the ducks' behaviour the speed at which you are able to reduce the temperature. Even at the end of their fourteen days in the brooder the heat is not turned off completely, and at this point we remove them from the brooder and take them to the old-fashioned hay boxes.

By the end of the month the first Chinese Geese will also be hatching. They hatch as easily as the ducklings and can be reared in the brooders in precisely the same way. When they are removed to the hay boxes they are separated from the duck and kept in groups of fifteen.

APRIL

The hay box is so called because there are two bags of loosely packed hay, placed on a wire tray about 10 in (25 cm) above the ducks' backs in the sleeping quarters. There is a wire floor which allows good circulation of air and lets the excreta fall through the floor, so that the ducklings stay clean and healthy. They are fed by a feed box on the side of the run and a fountain-type drinker which does not allow them to get into the water, so they do not get too wet.

In the methods of rearing that we use at Folly Farm, we find a group of thirty ducklings ideal, both in the brooder and when they move to the hay box. It is important, when they move to the hay box, that you have sufficient birds to generate enough body heat for the warmth to be maintained by the loosely packed bags of hay just above their heads. Ducklings will grow extremely fast in these conditions and, depending on the weather outside, if they appear to be getting too warm at night, simply remove one of the hay sacks. They usually remain in the hay boxes until they are about five weeks old.

By early April, the heavy geese will have started to lay. As we have had only moderate success with these eggs in the incubator we like to place them under broody hens or a light breed of goose which has gone broody. If neither of these are available, we take a chance and use the incubator.

April is the busiest month at Folly Farm. This is the peak time for setting eggs, tending to the incubator and the growing young stock. Also by this time we are beginning to welcome a steady flow of visitors.

MAY

By the beginning of May we can assess how well the hatching has gone. If we wish to set a few more eggs of any particular breed which has not laid or hatched as well as expected, we will extend our hatching for a further week, but we prefer all our ducklings to be hatched by 1 June. May is also the month when we begin to move the young stock from the hay boxes to the natural conditions outside, and for the first time they have swimming water. We remove the birds in their groups of thirty to a house that has been littered down with clean, dry straw. We do this at night, and the next morning they are let out for their first experience of swimming on a small pool, and how they enjoy it! But on their first visit to the pool we make sure that the odd one doesn't stay too long and get cold. After about a quarter of an hour we drive them back on the bank. Once they have got over the excitement of their first 'dip', they seem to be able to get their feathers well oiled and have no problems thereafter.

By this time the eggs from the heavy geese in the incubator will have hatched. The Toulouse geese need special attention for they tend to be slow and lethargic, taking longer to gain their feet. But when sufficiently strong they can be removed to the brooder and raised in much the same way as other geese. We like to leave goslings hatched by foster mothers to be reared by them as naturally as possible.

JUNE

The ducks and geese stay on the small pool until they are well feathered, and it is at this stage that we change their food from chick crumbs to growers pellets which they have ad lib in their house. After a week, we also introduce wheat, starting off with a handful or two just along the shallows once a day, until they get used to it. Then we gradually increase the amount making sure that they have finished their ration in about a quarter of an hour.

June is a month when the sun can be quite warm, and it is very important indeed for ducks in their growing period to have plenty of shade. They can quite easily get sunstroke and they go 'off their legs'. If a duckling should be affected in this way, we take it into a cool house, out of the sun on plenty of dry straw, making sure it has sufficient food and water; usually in a week to ten days it is back on its feet. Also, as the weather warms up, if their water is static it will become green and unpleasant, so we change it at least once a week.

As June arrives, the amount of work and the routine in the incubator house starts to reduce. We turn the machines off no later than the first week in June and concentrate more on the growing stock, moving the first hatch onto the large lake where they will remain until they either go back into the breeding pens or are sold to other breeders. When they are on the holding lake, they have no housing whatsoever, being protected from predators by a fox-proof fence. By this time they are already accustomed to eating pellets and grain, and now they go on to the twice a day feeding. They soon get used to being on the larger lake with other birds and, although they may be reluctant to come up and feed for the first time or two, they are very quick to learn.

Grass is abundant in June and the geese enjoy the good grazing, but although the adult stock can survive on grass alone we feed grain to them throughout the year. Our young stock would never achieve the growth we require on grass alone.

JULY

Now we begin to select our birds for our future breeding programme. We pick out the birds that are well grown and well marked, check their wing tags so that we know which strain they are, and transfer them to separate pens. We also begin to sell birds to other breeders, making sure we do not sell off the birds required for our own future breeding programme.

This is an interesting month to note the growth and conformation of the various strains, but we will still be unable to identify the birds that are perfectly marked because at this stage they will be in what we call the eclipse, a trait which persists despite the selective breeding of our modern domestic species. When the wild variety, from which these breeds have originated, have laid their eggs, the males will lose their gay colouring for the rest of the summer period so as not to attract predators to the nest, not returning to their full colour until late October. The same applies to our domestic breeds. A few years of keen observation will help you to forecast the likely outcome of birds in the eclipse to full colour.

One advantage to keeping waterfowl over other poultry is that, whether in the eclipse or full colour, wet or fine weather, they are always creatures of beauty. As we are open all the year round our visitors can appreciate the waterfowl whatever the season.

AUGUST

Our climate is so unpredictable – it can be as dry as 1984 or as wet as 1985 – but I must say that when keeping ducks, I much prefer the latter. Our ducks enjoyed the cool weather in 1985, diving and splashing about in clean, fresh water which proved most entertaining to visitors, lifting some of the depression created by the lack of sunshine and making them realise that not every creature likes a hot dry summer. However, August is usually dry and it is of the greatest importance that the ducks have access to clean water to drink and in which to immerse their heads. Even if the ducks are on a lake, strong sunlight and warm weather will soon turn water stale and green. So, in these conditions, it is important that they should always have access to a tub of clean water, and also plenty of

Ducks and geese with visitors at Folly Farm.

Some other 'visitors' to Folly Farm

shade for them to be able to rest comfortably in the heat of the day. If the water gets particularly stale, as it did in 1976 and 1984, it would be unwise to feed them in the water as we usually do; instead, feed them at dusk on the bank. As we do not want to feed all the crows, jackdaws and sparrows we feed the stock as late at night as possible, when the 'thieves' have gone to roost. Ducks are most active at this time after hot weather and, as they are naturally night feeders anyway, we find it best to feed both grain and pellets once a day, late at night whilst hot, dry weather persists. We adopted this method of management on three previous dry summers and have found that the birds have maintained their body weight and sustained excellent health until the return to more normal climatic conditions.

August is also the month when we are called upon to appear at a few exhibitions. This gives an excellent opportunity to display our ducks to those people who are not yet aware of how attractive domestic ducks can be. Under these circumstances, we do not put them in show cages but make a circle of green plastic covered wire with a fibreglass pool and decorate with a few shrubs and trees to make it look as natural as possible. This gives the public an impression of how it could look in their own gardens or paddocks. Also, we explain how easy it is to keep ducks. It is no good breeding and trying to improve our breeds if we neglect the opportunity to promote the idea and spread interest in keeping waterfowl.

SEPTEMBER

We use September to check all our equipment such as the hay boxes, the coops and the runs. We bring them all into the dry and cope with the repairs; when this has been achieved they will all need to be creosoted both inside and out. We thoroughly clean and fumigate the incubator and wash down the brooders with a strong disinfectant solution. We make a thorough check on all the houses that hold our breeding stock, making sure there are no loose or rotting boards – with the approach of winter the fox will be getting hungrier and is very partial to duck! He will be quick to spot any weaknesses which may allow him to gain access to the occupants. It is also a busy month for visiting breeders, many of whom are keeping duck for the first time, so a good deal of time is spent discussing the breeds and their requirements and catching the birds to which they have taken a fancy.

New breeders seldom bring anything in which to transport their birds so we are constantly on the look-out for second-hand cardboard boxes which we unfold and store flat. When they have purchased their birds we select a box of the right size and re-assemble it, placing a little hay in the bottom to absorb any moisture, then we punch holes in all sides of the box allowing plenty of air to circulate. We place the duck, or ducks, inside and close the lid; they usually settle quietly and travel without undue stress. We find this is a satisfactory way of transporting waterfowl. If the weather is hot we try to persuade the new owners to place the box on the back seat of the car rather than in the boot. Some car boots are almost air tight and if it is a sunny day the boot can become far too hot. If the boot is left slightly ajar, with the intention of letting the air circulate, in some vehicles this only draws in exhaust fumes and, on a few occasions we have witnessed people opening the boot of their car to find the poor creatures either dead from heat or gassed by the fumes. We are very careful to warn people of what might happen to birds travelling in the boot of their car.

OCTOBER

This is an exciting month with all our birds approaching full colour, so now we are able to assess just how successful our efforts have been in sustaining the breeds and even improving them a little. The result can be disappointing; it happens to all of us from time to time, and thank goodness we don't know all the answers as this teaches us to respect nature's ways. You must not be downhearted, but try to trace where there could have been improvements. When setbacks occur they should act as a spur to achieve a better result next time.

If we think we have bred a possible winner and we decide to enter it for a show, we know it must be presented professionally. It is not enough just to present the bird in top condition. We lift it from the run the day before, washing off any splashes of mud under a running tap, thoroughly clean the legs, scrubbing them if necessary, then wipe off the bill and place the bird indoors on some clean straw. We make sure the drinking bowl is spotless and feed the duck with wheat, as mash is far too messy. Next morning, we lift into a well-ventilated box, littered with dry straw. This is an operation over which we do not like to be rushed; a great deal of concentration is required and interruptions are not welcome at this stage. It is not easy to get the birds from the pools to the show pens in spotless condition, especially if it is raining or snowing when the birds are caught. Beginners should know that no matter how good the bird they are showing, if it is not in good condition and spotlessly clean it will considerably reduce its chances of winning its class.

NOVEMBER

This is the time of the year when I like to spend a little time watching the birds I have selected for the breeding pen, making sure that every bird is sound and their final colouring is the best I have. If any changes are to be made, this is the time to do it. At Folly Farm we have a succession of pools with water cascading from one to the next down the sloping terrain. During this month we try to find time for maintenance and improvement to the pools. Ducks have a habit of dabbling around the outside of the pools, which can destroy the banks. We have found the best way to avoid this is to peg small meshed wire firmly onto the bank of the pool. This will stop the ducks digging into the side and grass will begin to grow beneath the wire so that it soon becomes a clean grassy slope. This sort of work is expensive and time-consuming but each year we have managed to maintain a little progress. It is now pleasing to see the trees and shrubs that we planted three, four and five years ago beginning to provide more shelter and shade for the waterfowl and adding considerably to the beauty of the landscape. When planting trees, make sure that they are protected from geese who will strip the bark off young trees and destroy them in no time.

Small meshed wire protects the pool banks from damage by dabbling ducks.

After the first frosts, grazing is reduced in quantity and quality and additional corn has to be fed to keep geese in first class condition.

DECEMBER

We must be prepared for a deep snow in this month, and this will spell hard work for the stockman. It is likely that we will have to dig away the snow from the front of the houses to allow the ducks to come out to feed and water. Sometimes the snow has drifted higher than the fences and the birds could, if they wished, become mixed, but by this time they all know their own houses and do not stray too far from them. No matter how cold it is they still like to come out. The worst problem with snow is that you have to be prepared to spend the whole day simply making sure that every bird has food and water. Although this is the worst condition we have to cope with, so far every year we seem to manage and the birds survive fit and healthy.

By the end of the month when Christmas comes, we look back over the year and analyse our successes and failures. From this annual experience we can apply our accumulated knowledge with the hope of making improvements in the coming year.

11 Looking to the Future

There is still plenty of opportunity for new breeders to try their hand at breeding top quality stock. You will, of course, have to pay a higher price for the best birds in the first instance, but after that, the costs involved are precisely the same in catering for the needs of a valuable pure bred, as they are for a cross-bred of little value. Although breeding waterfowl may be a hobby, we would expect the sale of our surplus birds to, at least, recover the costs of breeding them. It is absolutely essential that we have a sound financial basis on which our breeders can rely and receive a just reward for their hard work and expertise, because this is the way to stimulate progress. Some people may consider it purely a hobby and may not wish to look at the financial side, but they should also remember that there are good breeders who cannot afford to lose money on every bird they produce and it would be a great shame if they could not afford to take part in the future breeding programme.

It is hoped that the newcomer will opt for one of the quality pure breeds. He should be warned that ducks lay an awful lot of eggs, and he must avoid the temptation, with his first flush of enthusiasm, to hatch too many. It takes some years to become established as a breeder and he cannot possibly expect to sell many pairs in the first year's breeding. He should be very careful not to produce more than he is able to sell or he will find his surplus stock taking up too much room and also costing far too much in feed. Faced with this problem, the best remedy is to kill them off for the table.

Some new breeders, faced with these circumstances, will either sell the birds off cheaply or even give them away, thinking they are doing somebody a good turn. They do not realise that the undervaluing of stock must be avoided, for this surely represents the greatest threat to, and inhibits the future breeding of pure bred waterfowl. New breeders who wish to play a useful and progressive part in the future must make every effort to improve and breed good stock and make sure they are sold at sensible prices. Remember that the costs must not only cover food, but also the membership of the breed's society, show expenses and the disposal of the surplus stock which do not come up to standard. Of course, you can always sell off these substandard birds as pets or table fowl, whichever you think best. It is most important, however, that you should never associate a pure bred name with the substandard duck that you are selling, as this is very damaging for the image of pure breeds and very misleading for the buyer.

A show is the ideal place to introduce the general public to breeding waterfowl. The words 'waterfowl show' or 'poultry show', at present used to advertise the event, do little to capture the interest of outsiders – the very people we wish to attract. They may well assume that it is a commercial activity, promoting commercial birds for meat or egg production. The word 'exhibition' is much more appropriate and coupled with a further explanation such as 'pure' or 'rare breeds', is much more likely to draw the attention of the public. After all, most of our breeds are well represented at these shows and ready to be viewed by the public, if only they were made more aware of what they were going to see.

So, exhibiting stock will play a very important part in attracting new breeders and improving our breeding stock for the future. The British Waterfowl Association is very aware of this and has recently established five regional shows, strategically placed in various parts of the country, so that as many breeders as possible will have the opportunity to exhibit their stock. Judges at regional shows will be from an approved list and have qualifications which meet with the approval of the British Waterfowl Association.

The East of England Poultry Club, who organise the first of the season's B.W.A. regional shows, were the first to adopt this method of billing, calling it 'The East of England Autumn Exhibition'. It proved highly successful and is growing from strength to strength. Once the public see waterfowl on show, their interest is maintained by the birds' natural appeal. I firmly believe that shows of the future will not only be there to provide sport for those with a competitive spirit but will also play a very important role in attracting new participants to the breeding and showing of waterfowl.

From the information that I glean from the increasing number of overseas visitors who visit Folly Farm every year, the upsurge in interest of keeping and breeding waterfowl is world-wide. I have spent many happy hours discussing the finer points of our British stock and comparing them to similar species in other parts of the world. We have made many interesting aquaintances and strong friendships have been forged. This will, I'm sure, be the pattern of the future and in the next few years domestic waterfowl breeding, and the British Waterfowl Association in particular, will have much closer links with its overseas counterparts, achieving an international flavour and hopefully, continuing to gain popularity on a world-wide scale.

I hope that within the pages of this book will be found inspiration and encouragement to all those who would like to become better acquainted with this absorbing hobby, so that they too may taste the joys that I have experienced in my lifetime in the keeping and breeding of domestic waterfowl.

Further Reading

British Waterfowl Standards (Butterworths in association with the British Waterfowl Association, 1982).

Grow, Oscar, *Modern Waterfowl Management and Breeding Guide* (The American Bantam Association, 1972).

Sheraw, Professor Darrel, *Successful Duck and Goose Raising* (Stomberg Publishing Company, 1975).

The Call Duck Breed Book (The American Bantam Association, 1983).

Soames, Barbara, *Keeping Domestic Geese* (Blandford, 1980).

Useful Addresses

The British Waterfowl Association
Secretary: Mrs Roz Taylor, Gill Cottage, New Gill, Bishopdale,
Leyburn, North Yorks DL8 3TQ

Fancy Fowl magazine
Crondall Cottage, Highclere, Newbury, Berkshire

The Poultry Club of Great Britain
Secretary: Liz Aubrey-Fletcher, Home Farmhouse, Fair Cross,
Stratfield Saye, Reading, Berkshire RG7 2BT

The Rare Breeds Survival Trust
Fourth Street, N.A.C., Stoneleigh, Kenilworth,
Warwickshire CV8 2LG.

Index

Page references for illustrations are indicated by italic type.

Index